No Sweat Projects

Hairy Science

PLANET DEXTER®

by Jess Brallier
Illustrated by Bob Staake

FEB 0 6 2001

Copyright © 2000 by Planet Dexter.
Illustrations copyright © 2000 by Bob Staake.
All rights reserved.

Written by Jess M. Brallier.
Designed by MKR Design/Pat Sweeney.
Cover design by Sammy Yuen snd Pat Sweeney.

Published by Planet Dexter, a division of Penguin Putnam Books for Young
Readers, New York.
PLANET DEXTER and the PLANET DEXTER logo are registered trademarks
of Penguin Putnam Inc.
Printed in the United States. Published simultaneously in Canada.

Library of Congress Cataloging-in Publication Data is available.

ISBN 0-448-44087-3 (pb.) A B C D E F G H I J
ISBN 0-448-44095-4 (GB) A B C D E F G H I J

Many of the designations used by manufacturers and sellers to distin-
guish their products are claimed as trademarks. Where those designations
appear in this book and Planet Dexter was aware of a trademark claim,
the designations have been printed with initial capital letters.

And Now a Message from Our Corporate Lawyer:

"Neither the Publisher nor the Author shall be liable for any damage that may be caused or sustained as a result of conducting any of the activities in this book without specifically following instructions, conducting the activities without proper supervision, or ignoring the cautions contained in the book."

A Hairy Guide

Introduction   6

Your Reports and Projects   7

But Why "Hair"?   8

Think Clearly   10

Getting Started   12

    Sample Research! 15

Getting Your Dexters in a Row   31

Hairy Experiments and Activities   34

    Hair's Looking at You! 34

    Compare the Hair 38

    Hair Appeal 41

    Hairy Warmth 42

    The Hairy Houdini 44

    I Spy . . . Nothing 46

    Har de Har Hair 48

    Hairy Crossword Puzzle 50

Really Helpful Stuff!   52

    Hair-y Anatomy 52
    Hair Anatomy 53
    Magnified Hair 54
    Crossword Puzzle Template 55
    Arctic Scene 56
    Plains Scene 57
    City Scene 58

Fun Stuff   59

"Hairy" People to Interview   61

Additional Sources   62

ACK!

Introduction

So why does this book exist?

- Need to do a science report or project?

- Looking for a subject that's really interesting and fun?

- Searching for an idea that'll impress your teacher and amaze your classmates?

- Need a subject that you know really well?

- Hoping to spend very little, or no, money?

- Are you running out of time?

This book is the answer to all those questions.

What this book will not do is your schoolwork. This book gives you ideas and illustrations you can copy, and it even gets you started on your **research**. But you have to do your own work. And guess what? It's going to be fun.

Hello!

Research is gathering information for your project.

So? What kind of science project do you need to do?

Reports and Projects Key

Oral Report

Group Activity

Written Report

Exhibit Project

As a student, you may be told to:

or Write a two-page or a five-page report.

Present a three-minute oral report to the class.

or Write a three-page report *and* present a three-minute oral report to the class.

or Write a three-page report *and* make a poster to be placed in the school cafeteria for parents' evening.

or Work with three classmates to do a written report *and* present something extra before the whole class.

or Present an oral report and use stuff like handouts, posters, etc.

Luckily, **Hairy Science** is perfect for any of these. As you use this book, it will tell you how ideas can be used for different types of reports and projects.

But Why "Hair"?

Science! Think about all the science in your hair!

Like:

Anatomy This branch of science studies how living things are built. What's hair made of? How's it attached to your body? Is it alive or dead? Can something dead grow? Is your hair just like the president's hair? Can a hair solve a murder case?

Zoology This branch of science studies animals—"zoo" is short for "zoological park." What if you were some dog or some bird rather than some person? Then your hair would be furry or feathery. What do hair, feathers, and fur have in common? What makes them different?

Genetics This branch of science explains why every living thing resembles its parents. Why do you have the hair you have? Why's your hair black and not blond? Curly and not straight? Why's your dad bald? Will you be?

Evolution This branch of science studies changes in living things, changes that sometimes take millions of years. Why do you have hair and not fur? Why do you have hair where you have it? (Why do you take your head in for a haircut and not your foot or elbow?) Why were prehistoric humans hairier? And one day will humans have hairless bodies?

OK, OK, but there's a bit of science in most everything. Why "hair"?

Familiarity!

Everybody knows hair! Hair's the scientific thing that people go most goofy over. They iron it if it's frizzy and perm it if it's straight. They sometimes change its color and sometimes shave it off. They hide it one day and show it off the next. See?—people find hair really interesting. (The same can't be said of your elbow or earlobe.*)

Cost!

Bringing a sample of hair to school costs nothing. Bringing a sample of a moon rocket to school costs about $9.8 billion. This "free" stuff is of great appeal to parents.

Everybody in class has it!

No need to buy or pluck 20 or 30 tufts of hair. You can get the entire class involved without handing out anything. (The same can't be said of Peruvian jellyfish.)

Fashion!

With a hairy project, you have the chance to wear a weird wig (purple dreadlocks or a pink Mohawk) to school. This is known as "adding flavor and flair" (those are good things) to your project.

Your teacher!

It's always best to keep your teacher happy. What better way than a scientific exploration of the stuff that sits on top of your teacher's head?

*"Hair" was a huge Broadway hit musical from 1968 to 1972. "Ear" wasn't.

Think Clearly: A Top Ten List

1 Before you do anything else, even before you go to the bathroom, **figure out what your project is.** Is it a group project? A written report? Two pages or five pages? Oral? Two minutes or five minutes?

2 **What do *you* have to do?** If you're working in a group, figure out what *you* have to do. (Even though it might not always seem like it, teachers know who the slackers are in any group.)

3 Get started so you can get finished. Don't wait to start! You might get sick. You might get invited to a party. Stuff like this really happens.

Research: Allow five days, one hour per day.

Writing: For a **written** report, write two hours the first day. On the second day, rewrite (make it better!) for one hour. Take the next day off (you deserve it!). Rewrite for 30 minutes on the fourth day.

Practice: For an **oral** project, practice your presentation for three days. On the first day, practice it out loud behind closed doors two times, and once with a parent. On the second day, practice it once — with any changes from your parent — behind closed doors, once more for a parent, and one more time behind closed doors. On the third day, practice it one last time in private.

An exhibit  should be finished two days before it is due at school. Ask family members to check it out for a day. Any problems with it? Tape coming off? Anything breaking? This leaves you one day to fix it.

4 **Check the spelling on *all* written materials.**

5 **Has somebody, like a family member, checked over all your materials?** Sometimes a different set of eyes sees stuff you don't.

6 **Presenting tomorrow?** Get a good night's rest.

7 **Look one last time at your teacher's instructions.** Have you done *everything?*

8 Pack the night before. Is your exhibit big or fragile? Have you figured out a way to get it to school without wrecking it? Don't wait until the bus is outside honking its horn to figure out this packing stuff.

9 If you can, pee before class.

10 Your contribution to this Top Ten List (whatever we forgot):

_____

_____

Getting Started
(The Amazing Note Card)

For as long as you are a student, you'll be doing school projects.

For every project you must do one thing: collect information (also known as **research**). That's what a school project is all about: collecting information and presenting it to somebody. That somebody may be a teacher or other students.

So how do you collect all that information? With the amazing note card.

The Amazing Note Card

You can collect information about hair, for example, by finding a book (like this one), a magazine, or a newspaper; or **watching** a video or TV show; or **searching** the Internet; or **interviewing** somebody (see page 61).

" ARCHAEOLOGISTS BELIEVE THE DOLL, ONE OF THE MOST ELABORATE EVER FOUND IN A PYRAMID, WAS STUFFED WITH THE HAIR OF A TWO TON YAK!

Whenever you find something interesting on your topic, write it down on a note card. Be sure to write on the card where you found the information (for example: *Hairy Weirdness: My Life as a Rock 'n' Roll Star*, a book by Joe Drummer, school library). Keep creating these cards until you have a stack of them.

You've done all your research? Great! Now arrange the cards in some order that makes sense. For example, cards on the same topic should be together. One card should follow another for good reason.

Imagine that if you taped all the cards together, like a string of cut-out dolls, you'd have a very rough first draft of a paper or presentation. **Neat, eh?**

On the next 16 pages are 58 note card-like pieces of information to get you started.

Read through some of them. Go ahead, right now. (We'll wait.)

See how they don't read like a normal book or paper? That's because these note cards are just the beginning of your work. They are not yet in an order that makes any sense. They're still like a bunch of letters (Y, I, A, H, R) that have to be put in order so that they make up a word (HAIRY).

For example, if you just copied these note cards out of the book and handed them to your teacher, the teacher would say something like, "That's a good start. **NOW FINISH IT!** Go write a paper that makes sense. These are just stones. Build a wall with them."

If this is a group project, think about how to share this note card work. Maybe it's by topic. Someone can research the anatomy of hair. Someone else may do fur, feathers, and camouflage. Someone else may research how hair, feathers, and fur are used to attract mates. Someone else may find all the "extra" weird stuff on the history and fashion of hair, hair trivia, etc.

You can buy packs of note cards or cut paper into pieces about 3 inches high and 5 inches wide (use this paragraph's note card as a guide). Keep your note cards together with a paper clip, or, if you do a lot of research, a rubber band will work.

Really Helpful Hint:
Once you have a lot of note cards, you should figure out what information is really **important**, what is **somewhat important**, and what is **not important at all**. So review all your note cards and mark those that contain really important information. For example, over the next few pages we've marked those note cards that contain the really important information with an **exclamation mark**:

OK?

And, hey, good luck with your note cards!

Sample Research!

1 ❗ EVOLUTION. Hair is with us from the get-go. About four months before a baby is born, silky, colorless hair called lanugo grows on most of its body. The lanugo grows first on the face (a baby beard!). Then it grows on the rest of the body. Finally, about a month before birth, all the hair falls off.

2 ❗ EVOLUTION. Millions of years ago—and before the invention of winter coats and home heating systems— humans were probably born with much more hair.

3 ❗
ANATOMY and GENETICS. A hair follicle is a very small dent in the skin out of which hair grows. The kind of hair you have depends on the shape of your follicles. Curly hair grows from rectangular follicles. Wavy hair grows from oval ones. Straight hair grows from round ones. People usually have the same kind of hair as the rest of their family.

4

EVOLUTION and ZOOLOGY.
Humans have about the same number of hair follicles that chimpanzees do.

YOU KNOW, JUDGING FROM THAT NICE HEAD OF HAIR OF YOURS, DON'T YOU THINK WE MIGHT BE RELATED?

5

Your hair grows fastest between 10 and 11 a.m. and 4 and 6 p.m.. These differences in growth are very, very small.

Lots o' leopard spots!

6

At puberty, hormones (substances in your body that regulate growth) cause body hair to show up. Male hormones cause beards to grow.

7

Assume that millions of years ago, females found bearded men attractive. Nature was helping humans attract the opposite sex, just as it happens with other creatures. The male peacock—a large bird—has beautiful blue and green feathers that attract female peacocks. And the male lion has a great-looking mane.

8

ZOOLOGY. The color of bird feathers serves two purposes. One, to camouflage (to blend into one's surroundings so as not to be seen by enemies); and two, to make a bird look good (to attract a mate).

9 !

ZOOLOGY. Hair on animals (fur) is often used for camouflaging. For example, leopards—large wild cats of Africa and Asia—have light brown fur with black spots. These colors blend with the plants in which the leopards hide.

10 !

ZOOLOGY. Hair is one of the things that makes mammals mammals. Mammals are hairy animals such as cats, deer, humans, and elephants.

11 !

EVOLUTION. Although they live in the ocean, whales are mammals. Scientists think that whales were once really hairy. But since the hair slowed them down in the water, over time they lost all but a few tufts.

12

A really important thing about hair is that it insulates a creature. In cold climates, hair conserves the body's heat. In the tropics, hair does the opposite, protecting some tropical animals from the dangers of excessive sun.

13!

ZOOLOGY. Long hair, like a mane, protects an animal's neck from the teeth of its enemies. And hairy tails serve as fly swatters. A porcupine's stiff quills—formed from bunched-up hair—protect the animal from enemies. And cat whiskers have special nerves that respond quickly to touch.

Watch out, buddy!

14!

GENETICS.

Hair Color	Number of Hairs on Head
red	90,000
black	108,000
brown	110,000
blond	140,000

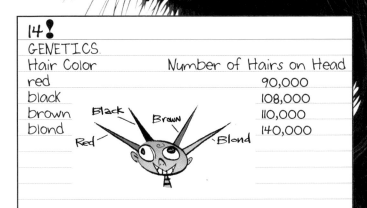

15

We all lose about 40 to 100 hairs a day. They just fall out normally, into our hair brushes, or onto the floor, or into whatever we're eating. Eyebrow and eyelash hair falls out before growing very long (never more than an inch).

16 ❗

Your hair grows about six inches a year, maybe more. Right now, about 90 percent of your hair is growing. The other 10 percent has stopped growing and is just taking it easy for about 100 days before it falls out.

17 ❗

Every person has about five million hairs on his or her entire body. Even if you subtract the hairs on your head, you're still talking roughly 4,900,000 hairs.

4,936,732...
4,936,733...

18

The darkest and thickest hairs on anyone's body are usually the eyelashes.

19!

Reach up and touch your hair. You just touched something dead. Except for a teensy spot at the place where each hair connects to your scalp, hair is completely dead. And yet, if washed now and then, hair is the kind of dead thing people like to look at and touch. (Weird!)

20

At one time, in addition to cutting hair, barbers pulled teeth and performed surgery. The red and white on a barber's pole stands for blood and bandages.

21!

ZOOLOGY and EVOLUTION.
Dogs shed when the weather gets warmer. The same is true of people. We shed more hair in the summer than we do in the winter, spring, or fall. Your hair also grows faster in summer than in winter.

22

Some people think that shaving hair makes it grow back thicker, or that cutting hair makes it grow faster. But the tiny part of the hair that's alive in the skin of your scalp can't tell if you're cutting or shaving the dead part of your hair. It just keeps growing in its normal way, no matter what you do.

Ahhh! I can't wait to shed my winter coat for the summer!

23

In ancient Greece, people sometimes cut off part or all of their hair and gave it to the gods whenever something really good (like getting married) or something really bad (like the death of a loved one) happened. Anyone wanting long hair had to hope for a boring life.

No more excitement, please! I'm losing all my hair!

24

The world record for long hair—at 13 feet, 10 1/2 inches—is held by Mata Jagdamba, of India. If she wanted to, she could play jump rope with her own hair.

25

Each hair contains some fat, which is what traps smells—like smoke or perfume—in it.

26

ANATOMY. Hair's made of:

50%	carbon	(what diamonds are made from)
20%	oxygen	(the same stuff you breathe)
17%	nitrogen	(what's used to make dynamite)
6%	hydrogen	(the same stuff used in rocket fuel)
5%	sulfur	(the same stuff that's in gunpowder)

And 2 % of random stuff

OLUTION. Assume that
llions of years ago humans
re covered with hair
hink of "cave men" in
ovies). Goosebumps are a
eminder of that. When
old, your hair stands on end
to create a trap for air and
provide a layer of insulation.
(Standing hair is what we call "goosebumps."

28 ‼

EVOLUTION and ZOOLOGY. When
you're scared, your hair stands on
end to make you look bigger so that
your enemy will be frightened of you.
Hair raising is an automatic gesture
triggered in many animals (dogs and
gorillas, for instance) when they are
frightened.

29

Without hair, you couldn't hear. Very small
hairs (you need a microscope to see them) in
the ears trigger nerves that allow us to
make sense of sound vibrations.

r nostrils contain hairs that are
ters. Without this hair, stuff like dust
d insects would go up your nose.

31

Without hair, your nose would drip snot constantly. There are itsy-bitsy hairs inside the back of your nose that sweep snot back into your throat where you swallow it away. When the hairs don't sweep—like when you have a cold—all the snot flows forward and out your nose.

32

ZOOLOGY. A cat uses her scratchy tongue as a kind of comb to lick herself clean. The tongue/comb pulls out a lot of fur, which the cat swallows and then coughs back up as a "hairball."

ACK!

33

Bezoars (BEE-zors) are the hairballs of goats, antelopes, llamas, and other cud-chewing animals. They are smooth and hard, like a marble. Some people find them attractive. England's Queen Elizabeth I (1533-1603) kept a bezoar framed in gold.

34 !

ANATOMY. Because hair is always falling out, police often find a suspect's hair at a crime scene. From one strand of hair, a crime lab can determine if the hair's owner is animal or human, what his or her race is, whether he's using certain drugs, etc.

35 !

ANATOMY. A crime lab can also tell if a hair fell or was pulled out, which part of the body it came from, whether it's real or from a wig of artificial hair, and if it has been dyed or bleached.

36 !

ZOOLOGY. Are there any bald mammals? Yep. The naked mole rat has no hair. It slithers around better in underground holes because of its baldness. Mexican hairless dogs are bred to be hairless, and certain cats are, too. But totally bald mammals are few and far between. Even an elephant is hairy (on its tail).

37 !

ANATOMY. You have a rain predictor growing on the top of your head. As the air becomes more and more moist (often a sign of rain to come), your hair lengthens by as much as 2 percent.

38 ❗

ANATOMY. In 1783 a hair hygrometer was invented to predict rain. A hair was tied to a needle. The needle moved as the hair lengthened and shortened, showing when to expect rain. A blond hair worked best.

Arrgghh! Me hair long.. no hurt me! Ha!

39

Hair screens out those rays of the sun that are harmful, makes great insulation (especially in winter), and cushions hits to the head. Ancient warriors, back when the most advanced weapon was a club, often grew their hair long for protection.

40 ❗

ANATOMY. Hair is made up of: the hair shaft (the part we see), the follicle (dent in the skin out of which hair grows), and the cortex. The cortex contains keratin, a protein, and the hair's colored pigment. Any keratin that dies joins the hair shaft (that's right!—your hair is dead keratin). A hair sits in the skin surrounded by sweat glands, layers of skin, muscles, and the sebaceous gland (which provides the hair's protective oils).

41 !

When a guy goes bald, the hair follicles in his head stop making the thick hair he's always had and begin making very fine, slow-growing, colorless (almost invisible) hair. So a bald man's not really bald, he just has hair that's really hard to see.

42 !

GENETICS. Whether a man goes bald has something to do with whether other people in his family have gone bald. But there's still no way to predict if the man will go bald, too.

43

Sometimes women lose hair, although they rarely go completely bald. Older women often notice that their hair is thinner, and some of them lose hair in patches.

44 !

GENETICS. Nature uses only two hair colors: red and brown. If your body uses no red and just a little brown, you're a blond. If your body uses a medium amount of brown and a little red, you'll have brown hair. Only red and you'll be a redhead. And black hair is actually just a very, very dark brown. Most of the world's population has dark brown hair.

45❗

GENETICS. Just because you're a blond kid doesn't mean you'll be a blond adult (though you might). On the other hand, if you're a red-haired kid, you'll almost certainly be a red-haired adult. The same goes for black hair. If you've got brown hair, it'll probably get darker, but it'll stay brown.

46❗

ZOOLOGY. (a) The only naturally blue hair you're ever going to see is on Kerry Blue Terriers. These dogs look black when born, but many of them grow up to be a dark blue-gray color. (b) The Duroc pig has naturally pink hair.

47

Pigment is the substance in the tissue of plants and animals that gives them their color.

NICE PIGMENT!

WHY, THANK YOU!

48❗

Hair is never gray. As people get older, they grow hair without pigment. The hair then appears white or almost clear. The next time you see a person whose hair appears gray—not white—take a close look. What you'll see is some white hair and some pigmented hair (dark brown, for example) on the same head. Together, they look gray, just as salt and pepper mixed together looks gray from a distance.

49

On average, if you're a white person, your hair will likely start going "gray" at age 34. If you're Asian, it should be a few years after that. If you're African-American, you probably won't see any gray until age 44.

50 ❗

ANATOMY. Why doesn't it hurt when you cut your hair and nails? Because your hair and nails are the only parts of your body that don't have nerves.

51

Hair and nails do not grow after death. That's only a myth. They only appear to grow because the flesh around them shrinks.

I DON'T KNOW HOW TO TELL YOU THIS, BUT YOUR NAILS LOOK DEAD!

52

Swimming pools are full of the chemical chlorine. If you stay in a pool for ten hours, the chlorine will melt away the outside of each hair shaft. This makes your hair dry and unmanageable. So rinse the water off as soon as you get out. (If you're a blond, chlorine may turn your hair green.)

53

If you look at the top of a left-handed baby's head, supposedly you can see that his or her hair swirls in a counterclockwise (leftward) direction. The hair on top of a right-handed baby's head supposedly swirls the other way, clockwise.

54

Every day tiny scales of skin rub off your body (including your scalp), float through the air, and land on the floor, furniture, lawn, or other people. People aren't bothered by this until they see these flakes on their shoulders—dandruff!

5

alf of all people between the ages of n and twenty have dandruff. After e thirty, the numbers drop.

56

Every three days your head makes a whole new set of skin cells. Unlike the rest of the dead skin cells on your body, the scalp cells get trapped in your hair. Sticky oil from the glands in your head sticks to these trapped cells. Then, dust joins the sticky, trapped skin cells. Welcome to dandruff! The good news is that most dandruff goes down the drain when you wash your hair.

Baaaa! I'm nice and toasty!

57

ZOOLOGY. Any animal with hair, feathers, or skin has dandruff. Yes, that includes parakeets, kittens, and walruses. The only difference is that it's called dander on nonhuman animals.

58 !

ZOOLOGY. Heat does not easily move through still air. Materials that trap air and keep it still are called insulators. Because sheep's wool is so curly, it traps air, makes a great insulator, and keeps you warm.

Getting Your Dexters in a Row

(Setting Your Priorities)

Well, What About Hair?

For your science project, are you going to present everything there is to know about hair? Or are you going to focus on hair and fur? Or on the anatomy of hair?

You've done some research by this point. What part of that information are you going to use? Or do you need to do more research?

Sample

Let's say your hairy science project will focus on human hair and how it is like, or not like, what grows out of animals. Go back to the note cards. Although the ones in this book are only a sampling, pretend they are all your note cards. Now mark those—go ahead, use a pencil, put a little check mark (✓) on each one—that you think you need to prepare a short oral report. Remember, your hairy science topic is "Human Hair versus Animal Fur and Feathers."

To check your selection against ours, see page 64.

Organizing All the Information

Be very clear as to what your presentation is about. You do this in the title to your exhibit , in the opening sentences of a paper , or in your first remarks in an oral report .

This will be helpful to your listeners and readers. But, more importantly, a clear understanding of what you're working on will be of great help to *you*. Let's pretend your presentation topic is "Human Hair versus Animal Fur and Feathers." You might want to write that down on a piece of paper and keep it in front of you.

Remember, in a paper or oral presentation, first clearly write what your topic is. And at the end, again write your topic in a summary.

Really Helpful Hints:

- This is a science project. Do not use goofy humor, superstitions, or rumors. Remember, your science teacher is looking for scientific information.

- Once you've picked the note cards from which you'll write a report, number them! Otherwise, dropping them might be a full-blown disaster.

- If you're doing an oral report and using note cards or sheets of paper, number them! That way, you won't worry about losing your place when you're in front of the class.

- If you need to refer to a written paper during your oral report, make doing that as easy as possible. Print it out in large, easy-to-read letters

LIKE THIS.

Hairy Experiments and Activities

Hair's Looking at You!

The first thing you need is a microscope. Maybe you can borrow one. Ask your teacher if you may use one of the school's microscopes. Be sure you know how to use it. Again, a teacher can often be of help. And every library has a book on how to use a **microscope**.

What you need.

- hair from several family members, friends, or pets
- plastic sandwich bag for each sample
- slides and cover slips (you'll get these when you get the microscope)
- water and eyedropper
- pen and paper or labels
- tweezers
- tissue paper or paper towel

What to do.

1 Collect hair samples from family, friends, or pets, and put each in a bag along with a small paper saying whose hair it is. You can either pluck the hairs out (if people give you permission!) or brush the person's or pet's hair and then remove the hairs from the brush.

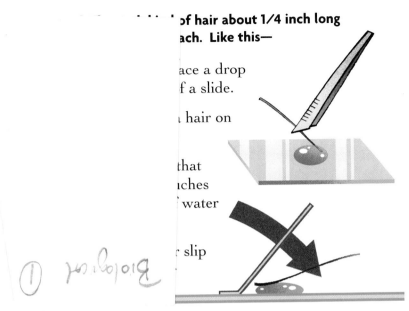

of hair about 1/4 inch long
ach. Like this—

ace a drop
f a slide.

hair on

that
ches
water

slip

- Blot up any extra water
  with a paper towel.

3 Look at the slide through the microscope.

4 Examine the other hair samples in the same way. Label
each slide before you make it. You may also wish to
draw pictures of the different samples.

What's going on.

ANATOMY. You should see three parts of each piece of
hair. 1) The outer layer is called the **cuticle**. You may
need to change the focus slightly to see the cuticle clearly.
2) Inside the cuticle is the **cortex**, full of keratin.
3) Inside the cortex is the **medulla** with its colored
pigment.

In older people, less pigment is produced, and so the hair
looks white. In general, brown hair tends to be thicker
than blond hair, and blond hair tends to be thicker than
red hair. If you are looking at dyed hair, it may have a
coat of different color on the outside of its shaft. You
may also see an undyed piece of hair near the hair's root.

Waaaa! I don't have a medulla!

ZOOLOGY. The sizes of the cuticle, cortex, and medulla are different in animals than in people. The hair of some animals (for example, sheep) has no medulla.

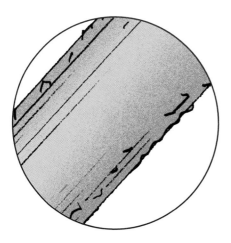

Hair under a high power (400 x) microscope: undyed blond hair (left) and dyed brown hair (right).

 Information from your microscope findings can be added to a paper. If you are writing on genetics, tell what you found from looking at blond, black, red, gray, and white hairs. If you are writing about animal fur, tell what you learned from viewing dog, cat, and hamster hair. Or maybe a local veterinarian can provide you with additional animal hair samples.

With a group project, you can divide up the work in many ways: one person can gather, bag, and label family and animal hairs; one person can do the actual microscope viewings; and one person can write and draw the findings.

Borrow a school microscope for an exhibit. Mount one hair for viewing. Also make a poster with descriptions—words and illustrations—of animal and human hairs. Then ask those looking at your exhibit to check off the correct hair on the following form. You can make many copies of the form, one for each person.

What Hair Is Being Viewed in the Microscope?

Hair	Check if correct
human (black hair)	_____
human (brown hair)	_____
human (blond hair)	_____
human (red hair)	_____
human (dyed hair)	_____
human (gray hair)	_____
animal (black hair)	_____
animal (black hair)	_____
sheep hair	_____

Viewer's name: _____

Compare the Hair

This activity can top off a hairy presentation in class. Or it makes for a near-perfect exhibit project.

What you need.

- Posterboard.

- Transparent tape.

- Hairs (or hairlike stuff) from lots of people, animals, or things, like—

a redhead

a blond

a dark-haired person

a gray-haired person

a straight-haired person

a curly-haired person

a doll

a fur coat, and

as many animals—including a bird's feather—as possible (dog, cat, hamster, horse, guinea pig, and so on). For these, a pet shop, grooming shop, or a veterinarian's office may be of help.

- Photos or illustrations of where each hair came from. If you don't have a photo of the dog, cut a photo of some other dog out of a magazine. If you don't have a photo of the gray-haired neighbor who shared a hair, cut a photo of some other gray-haired person out of a magazine. And so on.

What to do.

1 Across the top of the posterboard, put the name of your exhibit, like—

Compare the Hairs

2 Write a box of simple instruction, like—
Match each hair to whom or what it came from.

3 Tape each hair, one by one, onto the top half of the board and number each (1, 2, 3 . . .).

4 Tape each picture of the hair source onto the bottom half of the board and letter each (A, B, C . . .).

5 Create a simple answer sheet, like—

Hair	Hair Source
1	_____
2	_____
3	_____
4	_____
5	_____
6	_____
7	_____
8	_____

If you use this in class, keep the correct answers a secret and offer a small prize to whoever gets it right. If you use this as an exhibit, put the answers, in small letters, on the back of the answer sheet, like—

1-F, 2-C, 3-A, and so on.

Helpful Hint:

Don't you forget which hair came from where!

For a group project, divide up the tasks of collecting hair, creating the posterboard, and creating and copying the answer sheet.

WHOSE HAIR IS THIS?!

40

Hairy Appeal

In reading about hair, you discovered that hair, fur, and feathers often attract the opposite sex. Remember how important this is. If animals don't mate, they'll have no children, and soon there'll be no animals. **Oh No!**

Title a posterboard ("Hairy Appeal," for example) and then completely cover the board with photos of males, females, and couples. They can be birds or mammals, and be certain to include people.

Think of the peacock. Or a male and female cardinal (one's bright red, one's not). Imagine an animal that humans find ugly, like an opossum, and find a photo of two opossums together. There have been photos of moose who mistakenly fall in love with cows. Female lions go crazy for the mane on a male lion. Find photos of humans with outrageous hair styles, both from today and from years ago. Maybe you'll even find a woman with fancy hair wearing a hat of feathers and a fur coat. *(She's sure to appeal to somebody!)*

You can find photos and cut them out of old magazines (fashion and nature magazines are perfect) and old greeting cards (they often include photos of loving couples). Maybe you can even include photos discarded from your own family's photo files.

"Hairy Appeal" can be part of an exhibit. It can be one member's contribution to a group project. It can be shown during an oral presentation. And, if done on nonposter, normal-size paper, it can part of a written report.

Hairy Warmth

In your research, you discovered that sheep's wool or fur such as a rabbit's provides warmth. Here's your chance to prove it.

What you need.

- **An ice pack.** It should be as cold as possible. (This is the blueish stuff that stays cold for a long time after being in a freezer overnight.)

- **Cotton material.** You can use an old shirt, cotton gloves, or just a piece of fabric.

- **Wool material.** This can be a wool hat, gloves, or an old sweater.

- **Rubber kitchen gloves.**

- **Fur-lined winter gloves.**

What to do.

1 Ask for a volunteer.

2 Instruct the volunteer to pick up the ice pack with a bare hand; then do the same with each of the other items. Ask the volunteer to rank which item best protected the hand from the cold. "1" is the best protection, "5" provided the least protection.

3 Ask for more volunteers. Repeat the test.

4 Complete the chart below. Are your findings what you expected?

Hairy Warmth Chart

	Kept Hand Warm	A Little Warm	So-So	Pretty Cold	My Hand Froze
bare hand	_____	_____	_____	_____	_____
fur glove	_____	_____	_____	_____	_____
wool	_____	_____	_____	_____	_____
rubber glove	_____	_____	_____	_____	_____
cotton	_____	_____	_____	_____	_____

Because of the need to keep the ice pack cold, this is not the best of this book's activities for an exhibit. However, it makes for a great addition to an oral report.

For a group project, divide up the tasks of collecting the gloves and materials and creating, copying, and filling in the chart.

The Hairy Houdini

This fun activity involves a blindfold, a prop often used in magic, which is why this is called "The Hairy Houdini." (Many consider Harry Houdini, an American who lived from 1874 to 1926, to be the greatest magician ever.)

 This activity can be part of a hairy class presentation. Or, with some changes, it makes for a nearly perfect exhibit project.

What you need.

- At least four sandwich-sized plastic bags with a generous sampling (not one hair, but a tuft) of different hairs, like —

  human hair

  dog hair

  doll hair

  fur (check a thrift shop for an old fur stole; or turn a fur-lined glove inside out), and

  feathers (a pet shop, grooming shop, or a veterinarian's office may be of help).

- A blindfold (handkerchiefs work well).

What to do.

1 Ask for a volunteer.

2 Blindfold the volunteer.

3 Instruct the volunteer to put her hand into the bags, one by one (you hold the bags and guide the volunteer's hand), and ask her to guess from what sort of creature each hairy sample came.

For a "Hairy Houdini" exhibit, try the following:

1 Put each hairy sample in a brown paper bag and place the bags on a table. If you can have at least four samples, that's great.

2 Write what's in each bag (human hair, bird feathers, Barbie doll hair, and so on) on a card and place it upside down under the corresponding bag.

3 Use posterboard (a half-sheet will do) to title your exhibit, like—

Guess the Hair!

4 Also add to the posterboard simple instructions, like—

In each bag is hair or hairy-like stuff. Can you identify each source of hair or hairy-like stuff? Go ahead, reach in the bag. But remember—no peeking! The items are identified on the back of a card that sits under each bag.

5 Consider decorating the posterboard with hair or illustrations of hair, feathers, and fur.

For a group project, divide up the tasks of collecting hair, creating the posterboard, and getting the nonhair supplies.

I Spy . . . Nothing

As you discovered in your research, animal fur and feathers are often used for camouflage. Here's your chance to prove it.

What you need.

- **3 boxes**. Shoebox size is fine; they don't all have to be exactly the same.

- **3 scenes.** For example, a snow or arctic scene, a field of brown (African plains or a field of grain), and something like a photograph of an auto race or a colorful and busy city. You can illustrate these yourself, cut photographs out of magazines, or find samples in this book's "Really Helpful Stuff!" section (pages 52–58).

- **2 small stuffed animals** (like Beanie Babies). One should be white (like a polar bear or a white rabbit) and one should be light brown (like a brown rabbit or tiger).

What to do.

1 Place a different scene in each box.

2 Now, by placing each stuffed animal in the correct box, you can quickly demonstrate the effectiveness of camouflage. (This is also much cheaper than flying to an African animal reserve.) By putting one of your cute stuffed animals, like a rabbit, into the middle of a Las Vegas street, you'll amuse the class while clearly making your point.

WHERE'S THE BEAR?

THERE'S THE BEAR!

Neat Idea: You can make your boxes a lot better by adding dimension to them. For example, in the arctic scene, add cotton balls or soap flakes that look like snow. For the grassy scene, add real grass and twigs. And for a city scene, add some litter. These additions will further demonstrate how fur and feathers provide camouflage.

I Spy . . . Nothing makes for a great exhibit. Or, should you decide to do an oral presentation on something like "The Furs and Feathers of Camouflage," this is a fun activity to include.

Har de Har Hair

At the beginning or end of an oral presentation, it's sometimes fun to have your classmates goof off a little bit. Here are four simple and fun goofings.

1 **Hairy Challenge. Write the following information on a chalkboard:**

Hair Color	Number of Hairs on Head
red	90,000
black	108,000
brown	110,000
blond	140,000

Now, ask the class, "How many hours would it take a blond person to count his or her head hairs, counting one hair each second and not taking a break until finished? How about a redhead?"
(*Answer: blond, 39 hours; red, 25 hours.*)

2 **Dandruff Survey. Get a magnifying glass. (Ask your teacher for help.) Now you need volunteers with different colors of hair and hair length. This could be a problem. People are weird about dandruff—something that's really very normal.**

Use a magnifying glass to examine the scalp of each volunteer for dandruff (note cards 54, 55, 56, and 57 in the "Sample Research" section of this book). Describe the dandruff, out loud, to the class. Record different findings on the chalkboard. Does hair color or length cause more or less dandruff? (Probably not.)

3 How-to-Get-Rich-Quick Scheme. Assume that some wonderfully wealthy (but a bit goofy) person offers you either one dollar for each hair that grows on your head or one dime for each hair that grows on the rest of your body. Which offer should you take?

(Answer: Take the dime. Every person has about five million hairs on his or her body as a whole. Even if you subtract the hairs on your head [about 140,000], you're still talking roughly 4,900,000 hairs. You're looking at about $490,000 if you take the dimes; only $140,000 if you take the dollar offer.)

4 Dive, Barbie, Dive! When top swimmers race, they often shave their heads or wear bathing caps. Supposedly, a hairy head makes them swim a split second slower. Is there a way to prove this? Hmmm. What if you had a large fish tank of water? And what if you had two Barbie dolls? And what if you put a bathing cap on one (cut the finger tip off of a rubber glove) but not the other. And you dressed them the same. (Do you have two Barbie bathing suits? If not, just send your GI Joe action figure out of the room and do this with nude Barbies.) Get a watch. Get ready. Go!—drop both Barbies in the water, headfirst, at the same time. Who hits bottom first?

Hairy Crossword Puzzle

Ever make a crossword puzzle? A themed one? If not, hair's the perfect subject with which to create your first one.

Check out all your note cards (those from this book and others you've researched) for factual tidbits that suggest a one-word response. These are the hints that enable the puzzle player to guess the words that run across and up-and-down, intersecting with one another on the puzzle. Finding the tidbits of hairy information is fun; fitting and crisscrossing the words within the puzzle is even more fun.

To help get you started, a sample puzzle follows. And a blank puzzle is included in "Really Helpful Stuff!" (page 55).

At the end of an oral presentation, it can be fun to hand out copies of your puzzle to the class. Talk with your teacher about the crossword puzzle being a home-work assignment. (Just warn your classmates at the beginning of your presentation so that they'll pay close attention to what you have to say.)

This can also be part of an exhibit project. Make 50 copies of your puzzle for exhibit visitors to take and enjoy. They'll be thankful to learn so much about the stuff that sits on top of their heads. (Maybe they'll work on the puzzle while having their hair cut!)

Sample Crossword Puzzle

Across
1. Color with most hairs on head.
2. Deoxyribonucleic acid.
3. Season in which humans shed most.
4. What nose hairs sweep to the throat.
5. Hairball.
6. Hair so golden and curly.
7. Restores hair's pH balance.
8. Blue breed.

Down
5. Apparently hairless.
9. First bearded president.
10. Found in barn swallow nests.
11. When hairs stand on end.
12. Hair-based weather predictor.
14. Pink pig.
15. Blood-sucking hair dwellers.
16. Frenchmen's fashionable hairstyle of the 1600s.

Crossword grid answers:

- 9 Down: LINCOLN
- 10 Down: HORSEHAIR
- 1 Across: BLOND
- 2 Across: DNA
- 3 Across: SUMMER
- 11 Down: GOOSEBUMPS
- 4 Across: SNOT
- 12 Down: HYGROMETER
- 5 Down: BEZOAR / Across: BEZOAR
- 14 Down: DUROC
- 6 Across: GOLDILOCKS
- 15 Down: LICE
- 16 Down: BUN
- 7 Across: CONDITIONER
- 8 Across: KERRY TERRIER

Really Helpful Stuff!

These next seven pages are here for you to use as you wish. That's right!—Just go ahead and trace or photocopy these pages. You can then glue or tape them into a written presentation, copy and distribute them to your classmates, or attach them to a poster. It's up to you.

Hair-y Anatomy

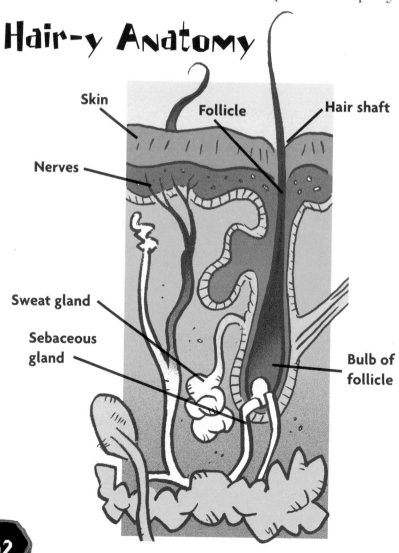

Skin

Follicle

Hair shaft

Nerves

Sweat gland

Sebaceous gland

Bulb of follicle

Hair Anatomy

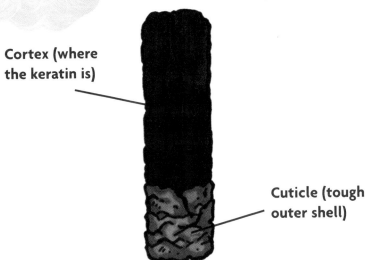

Cortex (where the keratin is)

Cuticle (tough outer shell)

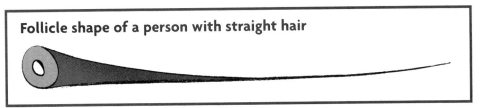

Follicle shape of a person with straight hair

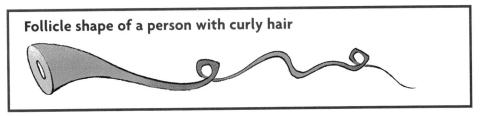

Follicle shape of a person with curly hair

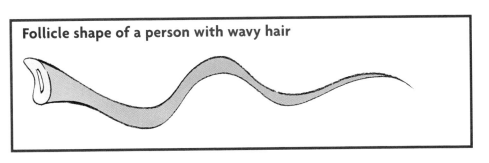

Follicle shape of a person with wavy hair

Magnified Hair

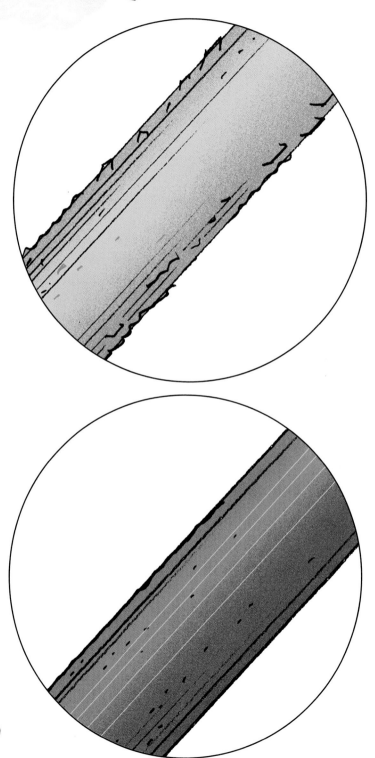

Crossword Puzzle Template

Arctic Scene

Plains Scene

City Scene

Fun Stuff

If you're doing an oral report, it's sometimes good to have a little fun with your listeners. Doing this actually helps them to pay attention. However, this kind of material is really not right for a written report.

Here's some amusing and informative stuff to drop into an oral report on hairy science:

"A celebrity is any well-known TV or movie star who looks like he spends more than two hours on his hair."—Steve Martin

Nit: Did you get a haircut?

Wit: No, I got all of them cut.

Find the hidden phrase (hold up a poster of this, or write it on the chalkboard).

HA / IR
HA / IR
HA / IR
HA / IR

Answer: splitting hairs

There's a flower called the "Hairy Beardtongue."

The longest beard on a man was over 17 feet long.

The longest recorded beard on a woman was a little over one foot long.

Really Helpful Hint:

Go back to the note cards on pages 15 to 30. Did you find some of them—like 4 and 16— more goofy than informative? These also can be a fun pause in your presentation. Ask if anyone knows why a barber's sign is red and white. Let them know. Ask if anyone knows at what time of the day hair grows fastest. And so on. Sometimes when you're doing research, you just don't know how you'll end up using the stuff you discover.

"So you're getting your hair cut tomorrow?"

"Yep. Hair today, gone tomorrow."

"Knock, knock."

"Who's there?"

"Hairy."

"Hairy who?"

"Hairy up and finish these awful jokes."

"Hairy" People to Interview

You will often learn more and have more fun interviewing people than you will doing any other type of research. When *you* actually interview people instead of reading books and magazines about those people, they are called *primary sources*. (Helpful Hint: teachers love it when you use primary sources.)

In researching hair, you may wish to interview the following people:

barber, hair stylist, or hair colorist

farmer

veterinarian

nurse

taxidermist

plumber (hair in the drain problems)

wig maker

grandparent

balding man

Additional Sources

Following are some of the books we've discovered while learning about hair. Most are available at your school or public library. You and a helpful librarian are likely to find other books that we missed. When searching the library, keep key words like the following in mind: "hair," "fur," "feathers," and "evolution."

Ardley, Bridgett, et al. *Skin, Hair, and Teeth.* Silver Burdett Press: Saddlebrook, New Jersey, 1988.

Jones, Charlotte Foltz. *Fingerprints and Talking Bones.* Delacorte Press: New York, 1997.

Parker, Steve. *Brain Surgery for Beginners and Other Minor Operations for Minors.* The Millbrook Press: Brookfield, Connecticut, 1993.

The Editors of Planet Dexter. *The Hairy Book: The (Uncut) Truth about the Weirdness of Hair.* Planet Dexter: New York, 1997.

Rudofsky, Bernard. *The Unfashionable Human Body.* Doubleday & Company: New York, 1971.

Sandeman, Anna. *Skin, Teeth & Hair.* Copper Beech Books: Providence, Rhode Island, 1996.

Schnurnberger, Lynn. *Let There Be Clothes.* Workman Publishing: New York, 1991.

Tomecek, Steve. *Simple Attractions: Phantastic Physical Phenomena.* W. H. Freeman and Company: New York, 1995.

And while at the library . . .

Search beyond books. Check out the magazine, newspaper, video, and microfilm catalogs.

Contact
The Hair Museum
815 West 23rd Street
Independence, Missouri (MO) 64055
phone: (816) 252-HAIR
(admission is charged)

Everything at the Hair Museum is "hair art," including more than 150 wreaths and 500 pieces of jewelry, all made of human hair. In the 1800s hair was used in ways that it is not used today. People mailed postcards with their own hair attached; necklace ornaments were made from a (usually deceased) loved one's hair; and hair wreaths, considered pieces of art, were hung on the wall.

Internet

Search for combinations of key words such as:

hair

scalp

baldness

fur

feathers

More No Sweat Science Projects!

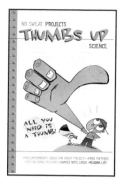

Thumbs Up Science

Confirming the notion that the human opposable thumb is as vital to the human race as is the brain, **Thumbs Up Science** guides the reader through scientific research, observation, and investigation. Along the way, the book provides a range of individual and group activities from thumb essentials (testing the opposable thumb and the secrets behind tendon-driven finger movements) to thumb fun (palmistry and thumbwrestling).

Shadowy Science

What are shadows? Why are they sometimes so big, sometimes so small, sometimes just the right size, and sometimes not there at all? How can shadows be used to tell time? Can there be shadows at night? Does a laser make a shadow? Great fun and great science, the shadow is a winning solution for most any science project.

Bouncing Science

Why are balls round and not square? What if they weren't? Why can you throw a baseball farther than a Ping-Pong ball but you can't throw a shot put farther than a baseball? Why can't you bowl on the beach? How do tennis players make the ball spin? How can a super ball bounce so high? Only science can answer these essential questions. This is the perfect subject for a well-rounded science project.

Suggestion Selection (refer to page 31)

If we were preparing a presentation of "Human Hair versus Animal Fur and Feathers," we would pull the following note cards out of those provided: 4, 8, 9, 10, 11, 13, 21, 28, 32, 33, 34, 36, 58, 59. Then we would quickly get back on the Internet and to the library because we need at least another 15 note cards on this specific aspect of hairy science.